AN INITIAL COURSE IN TROPICAL AGRICULTURE FOR THE STAFF OF CO-OPERATIVES

by

Peter Yeo, M.Sc.

Practical Action Publishing Ltd
25 Albert Street, Rugby, CV21 2SD,
Warwickshire, UK
www.practicalactionpublishing.com

First published 1976\Digitised 2013

ISBN 13: 9780903031394

ISBN Library Ebook: 9781780441627

Book DOI: http://dx.doi.org/10.3362/9781780441627

A catalogue record for this book is available from the British Library.

Since 1974, Practical Action Publishing (formerly Intermediate Technology Publications and ITDG Publishing) has published and disseminated books and information in support of international development work throughout the world. Practical Action Publishing is a trading name of Practical Action Publishing Ltd (Company Reg. No. 1159018), the wholly owned publishing company of Practical Action. Practical Action Publishing trades only in support of its parent charity objectives and any profits are covenanted back to Practical Action (Charity Reg. No. 247257, Group VAT Registration No. 880 9924 76).

CONTENTS

ACKNOWLEDGEMENTS

Financing for printing this series has been provided by the Ministry of Overseas Development. The Intermediate Technology Development Group gratefully acknowledge the Ministry's assistance.

Previous versions of this book have been tested over several years at the International Co-operative Training Centre, Loughborough, on courses for the Co-operative Officers and senior employees of Co-operative Societies. The participants in these courses have been generous in their advice. So have staff members of the Nottingham University School of Agriculture and of the I.C.A., and advisers on agriculture to the Ministry of Overseas Development in London. The author and the publisher are grateful to all these people for their help.

INTRODUCTION

0.1 PURPOSES OF THE COURSE. If you are a Co-operative Officer or otherwise concerned with rural development programmes in a tropical country, then this course is meant for you, unless you are fortunate enough to have had a training in agriculture already. Anyone working in rural development and particularly with Agricultural Co-operatives should know the language of scientific agriculture. He ought to know the purpose of the things Co-operatives sell to farmers. He ought to know in outline the conditions under which the products Co-operatives market for farmers can flourish. He needs to be able to understand the explanations given by agricultural experts. The idea that "agriculture is the job of another department and, therefore, does not concern the Co-operative Officer" is indefensible. Civil servants may try to think according to bureaucratic divisions but the farmers do not. They want co-ordinated help with raising their standard of living.

0.2 ITS LIMITATIONS. The course should help you to co-operate better with people who know more about agriculture than you do. Obviously, it doesn't make you an agricultural expert. It does not equip you to argue with trained agriculturalists about agriculture. It certainly doesn't justify any feeling of superiority over those who make their living by using traditional methods of agriculture. Most traditional systems are well adapted to the traditional environment. Population increase or the availability of modern techniques or markets may make changes seem desirable but great caution and expertise is needed to locate the best way of adapting to these developments. The traditional farmers can often teach the "experts".

0.3 WORKING METHOD. Please study each paragraph and answer the questions printed below each paragraph on a separate sheet of paper. (It is better not to write your answers in the book as you may want to go over the questions again later). Check your answers by turning the page and reading the correct answer overleaf. Go on if you have answered correctly; re-read if you have answered wrongly.

At the end of each part there is a Progress Test. Answer this test without referring back to the main text. Check your answers by reference to the answers which follow each Progress Test. If you have made more than two mistakes, study the whole part again. If you have made two mistakes or less, re-read only the paragraphs of the text covering the questions you have got wrong. This should be easy to find, as the relevant paragraph number is given after each answer. When you are completely satisfied with your performance on the Progress Test, go on to the next Part.

This is a demanding way of study in which there is no short cut and no diversions. However, it allows you to work at your own pace and in your own time. If you follow the procedure described above you should end up by knowing most of the facts listed. However, you won't remember them for long unless you take every opportunity to find out how they work out in practice, in the farming around you. In other words, your hard work on this text will have been wasted unless you use it as a basis for finding out more.

0.4 FURTHER EXPLANATION of all matters mentioned here can be found in *Tropical Agriculture* by Gordon Wrigley (pub. Faber & Faber). One of the several other good books on the subject is *Agriculture in the Tropics* by Webster and Wilson (pub. Longman).

Use the best book on Tropical Agriculture you can obtain to learn more, especially about the crops your Co-operatives handle. You need more information about the particular crops which concern you than it has been possible to include in this course.

Part 1

SOIL AND THE NATURAL ENVIRONMENT

1.1 ORGANIC MATTER. Good soil is alive. Tiny creatures within it are always active and a large part of good farming consists in encouraging useful activity and discouraging what is harmful. The small plants and animals in the soil (most of them too small to be seen by the naked eye), and their rotting remains, are called ORGANIC MATTER or HUMUS. This tends to be concentrated in the top few inches of the soil. Without it, the soil is dead and infertile. A layer of organic matter on the surface of the soil is called a MULCH.

Question

Top Soil is valuable because it contains

. . . .

(Answer by writing in the missing words)

1.2 MINERAL MATTER. Part of the soil comes from the break up of rocks. The resulting particles are classified according to size from gravel (over 2 mm in diameter) down through sand and silt to clay (less than 0.002 mm in diameter). Soil which has many of the bigger particles will have plenty of space between them through which air and water can pass. It is called LIGHT SOIL. If the tiny clay particles predominate the soil will be HEAVY. It will tend to be sticky and water and air will not pass through it easily. The plant foods get washed out of a light soil too easily and the water drains away too fast. The rain may not penetrate a heavy soil and it may be difficult to cultivate.

Questions

(a) Soil with too many big sand particles is called

. . . .

(b) Soil with many very small clay particles is called . . . .

1.3 SOIL STRUCTURE. The spaces between the particles are as important as the particles themselves. Plant roots penetrate where there is air.

ANSWERS

1.1 Organic Matter

1.2 (a) Light (b) Heavy

Most of the useful plants and animals in the soil need oxygen from the air. Several of the undesirable ones begin when air is excluded. A soil with a desirable amount of air space is said to have a good soil structure. Penetration by roots, the presence of organic matter and the right type of cultivation encourage good soil structure. The cultivation of heavy land when it is wet, over-cultivation which breaks the particles down too fine, pounding of exposed soil by heavy rains – any of these things make for bad soil structure. Extreme examples of bad soil structure occur when people, animals or vehicles kill all the plants and pack hard the top soil into an impenetrable layer by passing over it too often. If a similar impenetrable layer occurs some distance underground, it is called HARD PAN.

Question

A soil with a desirable amount of air space has a good . . . .

1.4 WATER. Plants take up food from the soil dissolved in water. The water clings round the soil particles absorbing what the plant needs from them into solution. The plant roots grow in between the particles and absorb the liquid. Without water, the plants cannot take up food. If too much water runs through the soil, the plant foods may be washed away and the soil is said to have been LEACHED. If too much water stays in the soil it becomes WATERLOGGED. Air is excluded and plant roots therefore cannot penetrate. Paddy rice is an exception here because it has air channels within its roots to bring air down from the surface through the water.

Question

Few roots can penetrate where there is too much

. . . .

1.5 WATER TABLE. Somewhere below most soils is a layer which is waterlogged. The top of this layer is called the WATER TABLE. Few plant roots can penetrate to it but the two or three feet of soil above it will have soaked up moderate amounts of water, and plants which have deep enough roots can draw water from this area. Rain or irrigation raises the level of the water table. Roots which are only temporarily flooded by this process should survive and benefit.

ANSWERS

1.3 Structure

1.4 Water

Question

Would a high water table always encourage growth?

1.6 SALINITY. If the water table comes close to the surface, water may soak up to the surface and be evaporated. The salts which it has dissolved in it may be left behind in excessive quantities and the resulting SALINE soil becomes infertile. This is a danger in irrigation projects and one reason for the paradox that drainage may be essential for successful irrigation.

Question

Irrigation can make soil saline without good .....

1.7 ORGANIC MATTER plays an essential part in water control. (If you have forgotten what Organic Matter is, turn back to 1.1.) Plant cover or a mulch protects the top soil from excessive pounding by the rain. Organic matter holds the soil together in small lumps between which water and air can penetrate. Organic matter holds some of the water that would otherwise run straight through before plants could use it.

Questions

(a) What is organic matter?

(b) Organic matter plays an essential part in .... control.

1.8 EROSION. Talking of water control leads on to Erosion. Wind and water may take away unprotected top soil when the particles get too fine. The faster the wind and water flow across the soil surface, the bigger the particles which can be carried away. Farmers should try to slow down these flows. That is why wind breaks are planted and why cultivation ridges should run across the slope rather than down it. (CONTOUR RIDGES). The previous paragraph should explain why another part of the remedy lies in maintaining organic matter content. The pounding down of the top soil described in 1.3 often leads to erosion.

Questions

(a) Erosion can be checked by .... the flow of wind and water across the soil surface.
(b) We should also try to keep up the amount of .... in the soil.

ANSWERS

1.5 No

1.6 Drainage

1.7 (a) Plant and animal matter in the soil and their rotting remains.
(b) Water

1.8 (a) Slowing
(b) Organic Matter

1.9 SUNLIGHT AND TEMPERATURE. Plants need light. It is an essential part of the process by which leaves draw in constituents from the air and turn them into plant food. This process is called PHOTOSYNTHESIS. It produces CARBOHYDRATES. This happens in the green parts of plants and you may have seen that plants deprived of light go white. Some crops like cotton and tobacco can use every bit of light available. With others, a factor working in the opposite direction can be important. Each crop has a temperature range within which it thrives best and, for many, the full sun of the tropics makes them too hot. Coffee and cocoa are examples of crops which may do better under shade. However, generally, the more nourishment a plant can draw from the soil, the more sun it can use. Thus fertiliser application may lessen the need for shade. Further information about carbohydrates will be given in 4.6.

Question

In the presence of light, carbohydrates are created in the green part of plants. What is this process called? . . . .

ANSWER

1.9 Photosynthesis

PROGRESS TEST ON PART I

Answer these questions in the space provided on the right hand side of the page. If dotted lines are provided, write your answer on them. The length of the lines gives you a very approximate guide to the length of the answer expected. Where no dotted lines are provided, write either TRUE or FALSE.

1. What is meant by ORGANIC MATTER or HUMUS in soil?

   . . . .

   . . . .

2. The top 25 centimetres of fertile soil must be kept free from biological activity.
3. Sand particles are bigger than clay particles.
4. The average size of soil particle is greater in heavy soil than in light soil.
5. Plant roots do not penetrate easily where there is no air.
6. Most of the useful plants and animals in the soil are active only where air can penetrate.
7. Cultivation of heavy land when it is wet usually improves the soil structure.
8. Without water, most plants cannot take up food from the soil.
9. If the spaces between the soil particles are completely filled with water, most plant roots cannot penetrate.
10. The lowest part of the waterlogged area in the soil is called the WATER TABLE.
11. Roots which are below the water table even for a short time die quickly.
12. Irrigation can leave the soil with too many salts in it.
13. Drainage can prevent soils becoming saline.
14. Organic matter improves soil structure.
15. What is the main purpose of CONTOUR RIDGES and wind breaks?

    . . . .

16. What is meant by PHOTOSYNTHESIS?

    . . . .

    . . . .

17. Cotton and tobacco can use every bit of light available. Coffee and Cocoa are examples of crops which may do better under shade.
18. Light is necessary for healthy plant growth.
19. Protection from full sunlight makes some plants grow better.
20. The more fertile the soil the less the need to shade the plants.

ANSWERS TO PROGRESS TEST ON PART 1

1. The plants and animals in the soil and their rotting remains. (1.1)
2. False. (A large part of good farming consists in encouraging biological activity in the soil. 1.1)
3. True. (1.2)
4. False. (1.2)
5. True. (1.3)
6. True. (1.3)
7. False. (1.3)
8. True. (1.4)
9. True. (1.4)
10. False. ("lowest" is wrong. It should be "top". (1.5)
11. False. (1.5)
12. True. (1.6)
13. True. (1.6)
14. True. (1.6)
15. To prevent erosion. (1.8)
16. The making of carbohydrates in green leaves in the presence of light. (1.9)
17. True. (1.9)
18. True. (1.9)
19. True. (1.9)
20. True. (1.9).

Part 2
FERTILISERS AND PLANT NOURISHMENT

2.1 NATURAL FERTILITY. Natural processes are the major sources of plant food. For a note on one of these processes turn back to what has already been said about PHOTOSYNTHESIS in 1.9. For four others read on:

Question

Is fertiliser the major source of plant food? ......

2.2 LEGUME is the name given to a family of plants which have bacteria on their roots. These bacteria can turn Nitrogen which is in the air into a form which plants can use. Beans and groundnuts are members of this family and so are clover and lucerne. Other plants with pods similar to those on the ones just mentioned are quite likely to be legumes, though not all pod bearing plants are legumes. Many cover crops which serve to keep down weeds and soil temperature between directly productive crops like rubber, are chosen because they are legumes. There are certain other organisms in the soil which can also make Nitrogen from the air available to plants in a similar manner. The process is often referred to as FIXING ATMOSPHERIC NITROGEN. It may stop if the soil is flooded for a long time or otherwise unsuitable for the bacterial activity.

Question

Legumes add .... to the soil.

2.3 TREES and other deep rooted plants may bring plant foods up from deep below the surface. The top soil benefits particularly when the leaves fall from the trees and rot. However, the fertility which trees create is likely to be quickly destroyed if the trees are cut and the soil is left uncovered.

Question

Do trees create permanent fertility? ....

ANSWERS

2.1 No

2.2 Nitrogen

2.3 No

2.4 ORGANIC MATTER has been mentioned many times already. It is an important part of the natural fertility of most soils. Plant and animal refuse add to it. Natural processes in the soil break it down to a form plants can use. If woody or unrotted organic matter is added to the soil it may be advisable to add Nitrogen at the same time.

Question

What should sometimes be added to the soil with woody or unrotted organic matter? . . . .

2.5 SHIFTING CULTIVATION. Traditionally, in most low-populated tropical countries, a system of shifting cultivation has been practised. Land which has been cropped for a few years and has lost its natural fertility is abandoned. It reverts to scrub or forest and its fertility builds up again. Many years later, someone comes and burns the natural vegetation before cultivating it again. The ash provides some extra natural fertiliser and the land may yield well again for a few years. Two main objections to this type of farming are that it is wasteful of land and that the natural fertility tends to get leached very quickly after the cover has been burned. (If you have forgotten what "leached" means, read 1.4 again.)

Question

What usually happens to natural fertility after the trees and bushes have been burnt? . . . .

2.6 MINERAL FERTILISERS. It may be profitable to add extra plant food. The major elements usually needed in quantity are Nitrogen (abbreviated "N"), Phosphate (abbreviated "P") and Potassium (abbreviated "K"). Note that as P is used as an abbreviation for Phosphate, K (latin "Kalium") has to be used for Potassium. Sometimes the name of a fertiliser may reveal what it contains. It shouldn't be difficult to recognise that something with "superphosphate" in the name contains P or with "nitrate" in the name contains N. Less obvious is that "ammonium", "ammonia" or "urea" will provide N and "basic slag" P. In most countries, each bag of fertiliser must be clearly labelled to show its contents. This is commonly expressed as units of N, P and K. For instance N_{10} P_{10} K_{15} indicates a fertiliser with more potash than nitrogen or phosphate.

Question

What does K stand for? . . . .

ANSWERS

2.4 Nitrogen

2.5 It is leached

2.6 Potassium

2.7 TRACE ELEMENTS. There are also certain Trace Elements or MICRONUTRIENTS of which only very small quantities are needed for plant health. Examples of these are Boron and Cobalt. In the case of shortage, trace elements can be applied to the soil or sprayed on to plant leaves. Excessive quantities of trace elements can be harmful.

Question

Trace elements are need in very ...... quantities.

2.8 SOIL ACIDITY. An important influence on the fertility of the soil is what is called its acidity. This is commonly measured on a pH scale on which low values indicate acidity and high values the reverse, which is called ALKALINITY. pH 7 is the neutral point between acidity and alkalinity. Soils outside the range pH 3.5 to 9.5 are not usually suitable for cultivation. The Nitrogen fixing bacteria (2.2) will not be active if the soil is too far from neutral. Each crop has its preferred pH level and there is a list of figures for many of them on page 53 of Wrigley's *Tropical Agriculture.* The pH level can be raised by adding lime or lowered by adding sulphur. Waterlogged or leached soil may be excessively acid.

Question

Is it true that low pH means high acidity? ....

2.9 WHAT KIND OF FERTILISER AND HOW MUCH? Fertiliser is very expensive and should only be applied if the extra income to the farmer which results is clearly greater than the cost of putting on the fertiliser. Different soil types lack different elements. Different crops need different elements. The cost of fertiliser and the selling price of the crop change from time to time. Farmers need to listen to the best advice they can get in order to find the most profitable type and quantity of fertiliser to apply.

Question

Should farmers apply fertiliser to all crops? ......

2.10 TIMING OF FERTILISER APPLICATION. Plant nutrient in the soil may be washed away by water coming down through the soil. You will

ANSWERS

2.7 Small

2.8 Yes

2.9 No

remember that this progress is called leaching. Alternatively, they may be transformed into chemical compounds which plants can't use. This is particularly likely to happen when fertiliser application has raised the level of nutrients in the soil above its natural level. For this reason, it is important to apply fertiliser as close to the time at which the plant will use it as possible. This presents problems to fertiliser suppliers. Everybody tends to want the fertiliser at the same brief periods of the year.

Question

The demand for fertiliser is likely to be much higher at some times of the year than at others. Re-read 2.10 to make sure you know why.

PROGRESS TEST ON PART 2

Answer these questions in the space provided on the right hand side of the page. If dotted lines are provided, write your answer on them. The length of the lines gives you a very approximate guide to the length of answer expected. Where no dotted lines are provided, write either TRUE or FALSE.

1. Artificial fertilisers are the major source of plant food.
2. What is the name given to the family of plants mentioned here as being able to FIX ATMOSPHERIC NITROGEN? . . . .
3. Will the fixing of atmospheric Nitrogen continue in soil that is flooded for a long time? . . . .
4. The fertility which trees create in a soil will always last long after they are cut.
5. Something else may need to be added to the soil at the same time as woody or unrotted organic matter. Name that "something else" . . . .
6. If trees and bushes are burned and the land left exposed, what is likely to happen to the natural fertiliser provided by the ash? . . . .
7. In fertiliser composition: What does P stand for? . . . .
8. What does K stand for? . . . .
9. What does N stand for? . . . .
10. Urea is included in fertilisers primarily as a source of Nitrogen.
11. Ammonium fertilisers are applied primarily as a source of Nitrogen.
12. Basic slag is applied to the soil primarily as a source of Phosphate.
13. The greater the quantity of trace elements applied the better.
14. In case of shortage, trace elements can be applied to the soil. What other method of application is also mentioned here? . . . .
15. What does a low value on a pH scale indicate? . . . .
16. What effect should the addition of sulphur have on the pH of ALKALINE soils? . . . .
17. Leached soils may be excessively acid.
18. What reason for adding lime to soils is mentioned here? . . . .
19. Fertiliser should not be applied unless the extra income which results is greater than the cost of putting on the fertilizer.
20. Farmers should always try to apply fertiliser as early as possible.

ANSWERS TO PROGRESS TEST ON PART 2

1. False. (Natural processes are the major source of plant food. 2.1).
2. Legumes. (2.2).
3. No. (2.2)
4. False. (2.3).
5. Nitrogen. (2.4).
6. It will be leached. (2.5).
7. Phosphate. (2.6).
8. Potassium. (2.6).
9. Nitrogen. (2.6).
10. True. (2.6).
11. True. (2.6).
12. True. (2.6).
13. False. (Excessive quantities of trace elements can be harmful. 2.7).
14. Spray on plant leaves. (2.7).
15. Acidity. (2.8).
16. Lower it or make it more acid. (2.8).
17. True. (2.8).
18. To reduce acidity or raise pH (2.8).
19. True. (2.9).
20. False. (It is important to apply fertiliser as close to the time when the plant will use it as possible. (2.10).

Part 3
CONTROLLING PESTS AND DISEASES

3.1 THE PARASITES. Crop yields can be badly reduced by weeds, animals, insects, bacteria, fungi and viruses. We shall use the word PARASITE here to include all these things when they are harmful to crops. This is a common way to use the word although biologists limit it to organisms which live in or on other organisms and draw their nourishment directly from the organism on which they live.

Question

The word. . . . . is used here for weeds, animals, insects, bacteria, fungi and viruses when they are harmful to crops.

3.2 CULTIVATION. Weeds can be controlled by hoeing or similar methods of turning the soil. The fact that this is a laborious process may not be important in a country where labour is cheap. What may matter more is that the hoeing may damage the crop or its roots. It may also damage the soil structure and encourage erosion. Nevertheless, this basic method of controlling weeds is likely to remain the most appropriate in many cases and, if the weeds go, some other parasites may also be reduced. This leads us into the next section.

Question

Is hoeing usually an appropriate way of controlling weeds if labour is cheap? . . . .

3.3 CLEANLINESS. Many parasites can only live on particular plants. Where this is the case, it may be possible to starve them out. This is the reasoning behind rules that there must be a gap between the time when a crop like cotton is cleared from the field and the time when the new cotton is planted. It is also a reason why it may be better to avoid growing the same crop in the same land year after year, but rather to use CROP ROTATION, which just means that the land is used for two or more crops one after the other. However, these types of parasite control cannot be

ANSWERS

3.1 Parasite

3.2 Yes

effective if types of weeds on which the parasites can live are left growing throughout. Such weeds are often called ALTERNATE HOSTS. Equally, they won't work if seed or diseased parts of the old crop are left behind to carry the disease through.

Question

Is it true that weeds and remains of the old crop can carry disease to new plants?

3.4 CLEAN SEED. It is important not to introduce disease with seed or other planted material. It is therefore sometimes necessary to treat the seed before it is planted or to persuade farmers to buy in planting material that is disease free, rather than carry over disease by using their own infected stock.

Question

The farmers' own planting material may carry . . . .

3.5 EXPLAINING REGULATIONS. Disease control is often enforced by government regulations. The regulations will not be obeyed unless the farmers are told the reasoning behind them and shown that it is in the general interest that individuals should be forced to obey them.

Question

Is it advisable to order farmers to obey rules they do not understand?

3.6 CHEMICALS. Parasites may be controlled by application of chemicals. These commonly come as sprays or dusts. Usually skill is needed to apply exactly the right quantity at exactly the right time. There is often danger that the chemicals may drift on to the wrong crop. Some chemicals like Sodium Chlorate may catch fire easily. Many are TOXIC (poisonous) and must not be breathed in or allowed in contact with the skin. If one adds that the chemical and/or the equipment for applying it may be expensive, it becomes clear why poor farmers do not make more use of chemicals for the control of parasites.

Question

What does toxic mean? . . . .

ANSWERS

3.3 Yes

3.4 Disease

3.5 No

3.6 Poisonous

3.7 TIMING. In attacking a parasite, it is often important to apply the chemical at exactly the point in its life cycle at which it is most vulnerable. However, some chemicals are PERSISTENT in that they remain effective for some time after application. The chances of catching the parasite at the right moment are thus higher with PERSISTENT chemicals.

Question

Chemicals which remain effective for some time after application are said to be . . . .

3.8 SYSTEMIC INSECTICIDES. Things used for killing insects are called INSECTICIDES. Some Insecticides are absorbed by the plants they are protecting, so that it becomes poisonous to insects which eat it. These are called SYSTEMIC INSECTICIDES.

Question

Insecticides which are absorbed into the plant they are protecting, so that it becomes poisonous to insects which eat it, are called . . . . insecticides.

3.9 WEED KILLERS. When chemicals are used to attack weeds, a great problem is to kill the weed without harming the crop. Some weedkillers are like Paraquat in that they are inactivated soon after contact with the earth so that land to which they have been applied is soon safe for crops. Others work on the green parts of plants but not on the woody parts. With care they can thus be applied under tree crops is they don't touch the leaves. Weed killers which can be applied among growing crops so as to kill weeds without damaging the crops are called SELECTIVE.

Question

What name is given to weedkillers which kill weeds without damaging the crop? . . . .

3.10 HORMONE WEEDKILLERS. One group of selective weed killers make use of the fact that narrow leafed plants like grass, maize or rice, absorb them less readily than broad leafed plants do. They make use of substances called HORMONES which cause rapid growth. The precisely right amount is applied so that the broad leafed weeds will take in so much hormone that they shoot up too fast and die while the narrow

ANSWERS

3.7 Persistent

3.8 Systemic

3.9 Selective

leafed crop absorbs less and is not harmed. Of course, if the crop is itself broad leafed like cotton, tobacco or most vegetables, these weedkillers cannot be used and they don't kill narrow leafed weeds like grass.

Question

Should the weedkillers described above be used to kill narrow leafed grass among broad leafed beans?

3.11 CROP BREEDING. Research stations try to breed varieties of a crop which are resistant to disease and, with some diseases such as Blackarm of cotton they have been successful. While on the subject of crop breeding it is important to mention that many of the new varieties of rice, maize and wheat are capable of producing very big harvests. However, most of them are only able to produce the higher yields if they receive sufficient water and fertiliser. They may thus be useless without irrigation and fertilisers.

Question

To derive benefit from new varieties of grain, irrigation and . . . . are often needed.

3.12 BIOLOGICAL CONTROL. Sometimes, a parasite can be controlled by introducing another parasite to attack it. For instance, "prickly pear" has been killed off in Australia by the introduction of the Cochineal insect. Parasites have been brought from Uganda to Kenya to control the coffee mealy-bug and from Trinidad to Fiji to control coconut scale.

Question

. . . . control means using one parasite to control another.

3.13 MISUSE OF INSECTICIDES. Insecticides may kill off some harmless insect which has previously been controlling a dangerous parasite. The insecticide may thus make the parasite worse. Another danger is that a parasite may develop resistance to the chemical which is supposed to control it.

Question

Insecticides can cause parasites to increase by killing off . . . . insects which have previously been attacking a dangerous parasite.

ANSWERS

3.10 No

3.11 Fertiliser

3.12 Biological

3.13 Harmless

PROGRESS TEST ON PART 3

Answer these questions in the space provided on the right hand side of the page. If dotted lines are provided, write your answer on them. The length of the line gives you a very approximate guide to the length of the answer expected. Where no dotted lines are provided, write either TRUE or FALSE.

1. Hoeing as a method of weed control takes a lot of labour. Write down two other possible objections to hoeing.
   (a) . . . .
   (b) . . . .
   . . . .
2. What is meant by an "Alternate Host"? . . . . . .
   . . . .
   . . . .
3. To control pests and disease, a farmer should try to get the new crop planted before the old is off the land.
4. How can crop rotation discourage parasites? . .
   . . . .
   . . . .
5. To control pests and disease, farmers should always try to use seeds or other planting material which they themselves have saved from last year's harvest.
6. What is said here that farmers must be told about disease control regulations, if they are to be expected to obey them? . . . .
   . . . .
   . . . .
   . . . .
7. What does TOXIC mean? . . . .
8. What particular danger is mentioned in connection with SODIUM CHLORATE? . . . .
   . . . .
9. The more crop protection chemical a farmer sprays on a crop against parasites the better.
10. It is always best to apply crop protection chemicals very early in the year.
11. What do you understand by the term PERSISTENT as applied to insecticides? . . . . .
    . . . .
12. What do you understand by the term SYSTEMIC as applied to insecticides? . . . .
    . . . .
    . . . .
    . . . .
13. Why is it said that it would be safe to plant crops in land which had recently been treated with PARAQUAT weed killer? . . . .
    . . . .
    . . . .

14. Some weedkillers can safely be applied under tree crops provided that none gets on the leaves.
15. HORMONE weed killers might well be used to kill grass under broad leafed crops.
16. HORMONE weed killers might well be used to kill broad leafed weeds under wheat or millet or other cereals.
17. The benefits from some new varieties of grain cannot be gained unless extra fertiliser and other inputs are supplied.
18. Why is it said here that parasites have been deliberately introduced into some countries?

. . . .

19. How is it said here that an insecticide might cause a pest to multiply? . . . .

. . . .

. . . .

ANSWERS TO PROGRESS TEST ON PART 3

1. a) Damage to crop or its roots.
   b) Damage to soil structure, causing erosion (3.2).
2. Weeds on which parasites can live even if the crop is cleared. (3.3).
3. False. (Parasites from the old crop may infect the new one) (3.3).
4. If a different crop is planted next year, parasites which cannot live on it will suffer (3.3).
5. False. (3.4).
6. The reasoning behind them, and that it is in the general interest that individuals should be forced to obey them. (3.5).
7. Poisonous (3.6).
8. It may catch fire easily (3.6).
9. False. (Skill is needed to apply exactly the right quantity) (3.6).
10. False. (Many chemicals should be applied at the point in the parasite's life cycle when it is most vulnerable) (3.7).
11. They remain effective for some time after application. (3.7).
12. Insecticides which are absorbed by the plant which they are protecting, so that it becomes poisonous to insects which eat it (3.8).
13. Paraquat becomes inactive soon after contact with earth (3.9).
14. True (3.10).
15. False. (The broad leafed crop would absorb more than the grass (3.10).
16. True. (3.10).
17. True. (3.11).
18. For Biological Control (3.12).
19. By killing off harmless insects which have previously been attacking a dangerous parasite (3.13).

Part 4
ANIMAL HUSBANDRY – I

4.1 INCREASING IMPORTANCE. There are not at present many tropical countries where much science is applied to animal management. However, as the number of people who can afford to buy animal products increases this situation is changing. It is therefore important to be able to understand something of the scientific background to the changes which are being made.

4.2 MANAGEMENT. The first paragraphs will deal with the management of animals as they now are and it is only later that GENETIC improvement, which is improvement of inherited qualities, will be discussed. This is deliberate since output of livestock products in most tropical countries could be greatly increased by simply looking after the existing types of animals better. Genetic improvements are often useless unless the standard of management is also improved.

Question

What does Genetic improvement mean? . . . .
. . . .

4.3 RUMINANT DIGESTION. Cows, sheep, goats and water buffalo are RUMINANTS. This means that they have compound stomachs. The first stomach is called the RUMEN. In it there are bacteria which can break down coarse material into a form which can be digested in the later parts of the digestive system. Ruminants can digest coarse food which would not be beneficial to other animals.

Question

What are animals, such as cows, with compound stomachs called? . . . .

4.4 DIGESTION IN BIRDS AND PIGS. Birds are not ruminants but they have a CROP in which some coarse material is broken down before it passes into the rest of the digestive system. Pigs can also digest some coarse material.

Question

Is it said here that pigs have crops? . . . .

ANSWERS

4.2 Improvement of inherited qualities.

4.3 Ruminants

4.4 No

4.5 FOOD ANALYSIS. Scientists can analyse animal feeds and quantify the elements in them which are useful to animals. Something is known of what animals need under various circumstances and scientific feeding consists in getting the right amount of nourishment for a given level of production into the animal as economically as possible. The main feed constituents are listed in the following paragraphs.

Question

Is it said here that animals should be given as much food as possible? . . . .

4.6 CARBOHYDRATES AND FATS. Animals need these for energy. Food analysts often express the amount of carbohydrates in cattle food in terms of "K" Calories of Energy (metabolisable or digestible energy). Maize, rice, bran, sweet potatoes and cassava are examples of feeds which are fairly rich in carbohydrates. MOLASSES is used as a carbohydrate supplement.

Question

What are Molasses and Maize rich in? . . . .

4.7 PROTEIN. These are particularly needed for body building and milk production. In food analysis, important considerations are the ratio of protein to starch and the digestibility of the protein. Rumen bacteria build up valuable proteins from feeds which would be useless to non-ruminants (see 4.3). In general, protein sources are expensive and it is wasteful to feed too much of them although it is also very important to be sure of feeding enough. Urea, cotton seed and ground-nut cakes are examples of feeds with good protein content.

Question

Is ground-nut cake likely to be good for milk production? . . . .

4.8 MINERALS. Important examples of necessary minerals are CALCIUM and PHOSPHORUS, which play a part in bone building and milk production. Animals can often be made much more productive by giving them cheap blocks containing minerals to lick.

Question

Is it always expensive to supply minerals to cattle?
. . . .

ANSWERS

4.5 No

4.6 Carbohydrate

4.7 Yes

4.8 No

4.9 BULK. An animal needs enough bulk of food to satisfy its appetite but it will not eat more. There may thus be no point in providing very concentrated foods which provide the necessary nutrients in a small volume without meeting the need for bulk. On the other hand, a high yielding cow will need concentrated food to give the necessary nutrients within the bulk of food it can eat.

Question

Would it be sensible to give a cow which is producing very little milk much concentrated food?

. . . .

4.10 SALT AND WATER. Many animals in the tropics would be far more profitable to their owners if they were provided with enough water and some common salt. The cost would in many cases be negligible.

Question

Provision of salt and water can be a cheap way of increasing profit from animals. Look back to 4.8 to find another cheap improvement.

ANSWER

4.9 No

PROGRESS TEST ON PART 4

Answer these questions in the space provided on the right hand side of the page. If dotted lines are provided, write your answer on them. The length of the line gives you a very approximate guide to the length of answer expected. Where no dotted lines are provided, write either TRUE or FALSE.

1. The improvement of inherited qualities in animals by breeding is called .... improvement. What is the missing word?
2. Improvement of inherited qualities in animals is often useless, unless the standard of management is also improved.
3. What name is given to cows and sheep because they have compound stomachs? ....
4. What does bacteria do in the first stomach of a cow? ....
....
5. Birds are not ruminants.
6. Birds can digest coarse material.
7. Scientific feeding is said here to consist in getting as much concentrated food into the animal as it will take.
8. "Food analysts often express the amount of .... in cattle food in terms of K calories of energy. What is the word which has been left out?
9. Urea, cotton seed and ground-nut cakes are examples of feeds with a good .... content. What is the word which has been left out? ....
10. What is the purpose of feeding Molasses, as given here? ....
11. Ruminants can derive digestible protein from food that non-ruminants cannot get much nourishment from.
12. In what are maize, rice, bran, sweet potatoes and cassava rich? ....
13. CALCIUM and PHOSPHORUS are listed here as being poisonous to animals even in small quantities.
14. Mineral licks are said to be expensive but worth using.
15. It is said here that it is usually best to feed the most concentrated food available to animals.
16. A cow which is producing a lot of milk has a special need for concentrated food.
17. It is said here that there are two common substances which are often very cheap but which, if supplied in adequate quantities, would make many animals kept in the tropics more profitable to their owners. Name the two substances.
   (a) ....
   (b) ....

ANSWERS TO PROGRESS TEST ON PART 4

1. Genetic (4.2).
2. True. (4.2).
3. Ruminants (4.3).
4. Break down coarse material into a form which can be digested in the later parts of the digestive system (4.3).
5. True. (4.4).
6. True. (4.4).
7. False. (the *right* amount and *economically*) (4.5).
8. Carbohydrates (4.6).
9. Protein (4.7).
10. As a carbohydrate supplement (4.6).
11. True. (4.7).
12. Carbohydrates (4.6).
13. False. (4.8).
14. False. (Mineral blocks are sometimes cheap) (4.8).
15. False. (4.9).
16. True. (4.9).
17. Salt and water (4.10).

Part 5
ANIMAL HUSBANDRY – II

5.1 PASTURE MANAGEMENT. Pasture is usually the best source for the major part of animal food for ruminants. If the land can be fenced, it may pay to plough it and resow with more productive grasses and legumes. Ideally the time and intensity of grazing should be controlled. Overgrazing makes pasture less productive and may even kill the grass and lead to erosion. On the other hand, pasture which is allowed to grow too long becomes coarse and low in nutritive value.

Question

Ruminants can usually get most of their food from well managed . . . .

5.2 MILK PRODUCTION. In most countries there is a time of year when natural pastures grow slowly or not at all. Milk production at this time becomes difficult and expensive. The output of milk thus varies with the seasons. If the demands of the market for liquid milk are met in the difficult times of year, there will be a surplus at the times when pasture grows well. The consequences for those who are involved in Dairy Co-operatives are important. They usually have to develop the production of things like milk powder, butter, or ghee to provide an outlet for the surplus milk.

Question

Can the amount of milk animals produce be kept constant through the year? . . . .

5.3 CONCENTRATES. A successful dairy industry also needs provision of concentrated feeds, particularly to keep up milk yields in the difficult time of year. It is worth noting that the residue after vegetable oil has been extracted from things like groundnut and cotton seed is a useful source of supplementary food for animals. This is one extra argument for extracting the oil within the country rather than exporting the whole seed. However, both this and the milk processing mentioned in the previous paragraph call for large inputs of capital.

ANSWERS

5.1 Pasture

5.2 No

Question

Can a good dairy industry be set up cheaply? . . . . .

5.4 FEEDING OF CALVES. In the interests of producing milk for sale, calves are taken from their mothers early. Two important matters arise. One is that the young animal must receive COLOSTRUM, which occurs in the milk of a mother after calving. The second is that very high standards of cleanliness are needed if young stock are to be reared successfully away from their mothers.

Question

Calves can be reared away from their mothers if conditions are clean and if they receive . . . .

5.5 DISEASE CONTROL. Many animal diseases are carried by insects such as ticks and tsetse fly. Spraying, dipping and bush clearance are aimed at keeping the insects off the cattle. Other animals may store infection which insects carry to livestock. For instance, tsetse fly can carry TRYPANOSOMIASIS from wild game to cattle. This may be a reason for killing off the game.

Question

Refer back to 3.3 to find words to describe wild animals which carry disease in this way. They are

. . . .

5.6 IMMUNITY. Some animals may be immune or resistant to certain diseases. They may be given this immunity by innoculation. They may acquire it by exposure to a mild attack of the disease as when young calves are exposed to East Coast Fever. They may also inherit it. In general, imported breeds have low resistance to local diseases, and this is one reason for crossing them with local breeds.

Question

Is it true that local breeds are usually better able to withstand local diseases than are imported breeds?

. . . .

5.7 SELECTION. Most herds can be improved by breeding only from the best animals in them. To do this successfully the farmer needs to be quite clear of what he wants. Is it more meat, more milk,

ANSWERS

5.3 No

5.4 Colostrum

5.5 Alternate Hosts

5.6 Yes

or better hides, or some combination of these? Which cuts of meat can be most profitable sold in his markets? His problems do not end here as you cannot identify high milk yield, economical conversion of food to flesh, or regularity of calving, merely by looking at the animal. Records of milk yield or other performance have to be kept for each animal.

Question

Can a good farmer identify a good animal by its appearance alone? . . . .

5.8 PROGENY TESTING. The best test of an animal's value for breeding is to observe the performance of its offspring *(progeny).* This is normally only worthwhile with the male animal. A bull which has been shown to be good by progeny testing is said to be PROVEN. It becomes very valuable.

Question

Is it true that a proven male animal is one which has been shown to produce good descendants? . . .

5.9 ARTIFICIAL INSEMINATION. Few farmers could afford to buy even one proven bull. That is why the practice of Artificial Insemination (A.I.) is so useful. A selection of bulls are kept at the A.I. station and their semen is collected and preserved. One ejaculation is sufficient to service many cows. When a cow comes on heat (is ready to mate) in the surrounding area, someone has to travel out with the semen and insert it. Good communications are needed to ensure that the semen arrives while the cow is still on heat.

Question

Is it true that good communications are needed for a good A.I. programme? . . . .

5.10 MIXED FARMING. In areas where both crop farming and stock farming are possible there is much to be said for integrating the two on the same farm. Cows provide their owner with a regular supply of high protein food and an income which is fairly constant through the year rather than being concentrated at harvest time. Animals can be fed at low cost on crop residues, which are what is left after the part which is directly useful

ANSWERS

5.7 No

5.8 Yes

5.9 Yes

to man has been taken. Finally, the manure is a valuable source of soil fertility.

Question

Should stock farming and crop farming always be separated? . . . .

5.11 SCOPE FOR IMPROVEMENT. Comparisons between continents show that there are enormous possibilities for improving the output from animals in most tropical countries. For instance, the average cow in Europe produces more than ten times as much milk per year as the average cow in Asia or Africa. Production of meat per head of beef cattle in North America is about six times as great as in Asia or Africa. Even after making allowance for climate and economic factors, the gap remains very large indeed. It is clearly very important to help farmers to use modern methods of keeping animals.

ANSWER

5.10 No

Answer these questions in the space provided on the right hand side of the page. If dotted lines are provided, write your answer on them. The length of the lines gives you a very approximate guide to the length of answer expected. Where no dotted lines are provided, write either TRUE or FALSE.

1. The longer pasture grass grows before the animals get to it, the better.
2. Grass is a crop which needs to be looked after carefully if it is to be very productive.
3. If the needs of the consumer for liquid milk are to be met throughout the year, there is likely to be more milk than this market can absorb at some times of year.
4. Why is it difficult to run a dairy industry without either good sources of supplementary feed or an outlet for surplus milk such as production of cheese or milk powder? . . . . . .
. . . .
. . . .
. . . .
. . . .
. . . .
5. What relevance has the feeding of cattle to the question of whether to export oil seeds or to extract the oil within the country where they are grown? . . . .
. . . .
6. What ingredient of a cow's milk is said here to be essential if calves are to be reared successfully away from their mothers? . . . .
7. What other aspect of the feeding of young calves is mentioned as being specially important?
. . . .
8. What reason for killing off wild game is given here? . . . .
. . . .
9. What reason against using pure bred imported animals in tropical countries is given here? . . .
. . . .
10. How might a mild attack of a disease like East Coast Fever be beneficial to a calf? . . . .
. . . .
11. A good farmer can identify high yielding cows by looking at the shape of their bodies.
12. Most herds of cattle in tropical countries could be improved by breeding without bringing in any animals from outside the herd.
13. What is said here to be the best test of an animal's value for breeding? . . . .
. . . .

14. What do you understand by a PROVEN BULL?

. . . .

. . . .

15. Why are good communications necessary for a successful ARTIFICIAL INSEMINATION programme? . . . .

. . . .

16. With good ARTIFICIAL INSEMINATION each bull services just one cow with each ejaculation.
17. A farmer's income from milk should be more regular than his income from crops.
18. Animals produce "food" for crops.
19. Animal food can be derived from crops without lessening the amount of plant food sold for human consumption.

Finally, please think carefully before tackling three final questions. If you get them wrong this INTRODUCTION TO AGRICULTURAL SCIENCE will have been worse than useless. It will have done great harm. (The answers to the questions lie in paragraph 0.2).

20. This course has equipped me to argue with trained agriculturalists about agriculture.
21. This course has equipped me to teach farmers about farming.
22. This course is aimed to make me better able to collaborate with trained agriculturalists and farmers.

ANSWERS TO PROGRESS TEST ON PART 5

1. False. (5.1).
2. True. (5.1).
3. True. (5.2).
4. The output of milk from cattle fed on natural grazing varies through the year with the seasons. If there is enough liquid milk in difficult times of year there will be a surplus when pasture grows well (5.2 and 5.3).
5. The residue after oil has been extracted is a good food for cattle (5.3).
6. Colostrum. (5.4).
7. Cleanliness (5.4).
8. Wild game may store infection which insects carry to cattle (5.5).
9. Low resistance to local disease (5.6).
10. To gain immunity (5.6).
11. False. (5.7).
12. True. (5.7).
13. To observe the performance of its progeny (5.8).
14. One which has been shown to be good by progeny testing (5.8).
15. To ensure that the semen arrives while the cow is ready to mate (5.9).
16. False. (5.9).
17. True. (5.10).
18. True. (Manure is a valuable source of fertility) (5.10).
19. True. (Animals can be fed on crop residues) (5.10).
20. False. (0.2).
21. False. (0.2).
22. True. (0.2).

www.ingramcontent.com/pod-product-compliance
Lightning Source LLC
Jackson TN
JSHW010854211224
75817JS00005B/131
9780903031394